EXCITING OUTDOOR CAREERS

PARK RANGER

DWAYNE HICKS

PowerKiDS press

Published in 2026 by The Rosen Publishing Group, Inc.
2544 Clinton Street, Buffalo, NY 14224

First Edition

Editor: Greg Roza
Book Design: Rachel Rising

Photo Credits: Cover, Vitalii Golovenko/Shutterstock.com; Cover, pp. 1, 3–24 klyaksun/Shutterstock.com; Cover, pp. 1, 3–24 Iurii Vlasenko/Shutterstock.com; p. 4 Erickson Stock/Shutterstock.com; p. 5 Joseph Sohm/Shutterstock.com; p. 6 Fotos593/Shutterstock.com; p. 7 Fit Ztudio/Shutterstock.com; p. 9 santdavalos/Shutterstock.com; p. 9 https://commons.wikimedia.org/wiki/File:Logo_of_the_United_States_National_Park_Service.svg; p. 10 Leonard Zhukovsky/Shutterstock.com; p. 11 David Watkins/Shutterstock.com; p. 12 wavebreakmedia/Shutterstock.com; p. 13 Orion Media Group/Shutterstock.com; p. 15 Dany Bejar photography/Shutterstock.com; p. 16 Gita Steponaviciute/Shutterstock.com; p. 17 anatoliy_gleb/Shutterstock.com; p. 18 https://commons.wikimedia.org/wiki/File:Harry_Yount_Berthoud_Pass,_Colorado.jpg; p. 19 https://commons.wikimedia.org/wiki/File:Ranger_Shelton_Johnson_addresses_guests_to_the_re-dedication_event._(b12beab6-2530-42f4-805f-cd73693bca08).jpg; p. 20 Chin Ting Chan/Shutterstock.com; p. 21 a katz/Shutterstock.com.

Some of the images in this book illustrate individuals who are models. The depictions do not imply actual situations or events.

Cataloging-in-Publication Data

Names: Hicks, Dwayne.
Title: Park ranger / Dwayne Hicks.
Description: Buffalo, New York : PowerKids Press, 2026. | Series: Exciting outdoor careers | Includes glossary and index.
Identifiers: ISBN 9781499457247 (pbk.) | ISBN 9781499457254 (library bound) | ISBN 9781499457261 (ebook)
Subjects: LCSH: Park rangers–Vocational guidance–Juvenile literature. | Park rangers–Juvenile literature.
Classification: LCC SB486.V62 H54 2026 | DDC 363.68023–dc23

Manufactured in China

CPSIA Compliance Information: Batch #QWPK26. For Further Information contact Rosen Publishing at 1-800-237-9932.

CONTENTS

PARK POLICE

If you like spending time outside whenever you can, you might want to think about becoming a park ranger. Rangers spend most of their working hours outside, even in bad weather. They work in national, state, and local parks. They work in national forests and **recreation** areas.

Park rangers are people who protect our nation's parks. This could mean arresting people who harm park property, or people who illegally hunt in parks. It could also mean putting out fires and identifying **invasive species**. However, park rangers do much more than that. They are teachers, emergency workers, **environmentalists**, and even maintenance workers.

OUTDOOR ADVICE

Park rangers often follow one of two career paths: law enforcement or educator. Many do both!

Park rangers may also be called forest rangers, park police officers, and environmental police officers.

SKILLS AND KNOWLEDGE

You can prepare for a career as a park ranger right now. Perhaps the best thing you can do to prepare is to stay physically fit. Rangers spend a lot of time walking forest trails or along creeks and shorelines. They are sometimes called upon for rescue missions.

Park rangers need good **communication** and leadership skills. They need to explain park rules clearly. They teach people about the dangers of wildlife and poisonous plants.

Park rangers are also good problem solvers. Quick-thinking rangers use their knowledge to keep tourists and hikers safe during times of danger.

OUTDOOR ADVICE

Park rangers must understand the local weather. They must let hikers know what the weather will be like for them. Sometimes they shut down a park when bad weather is on the way.

Rangers usually work a 40-hour week. However, some rangers are on call 24 hours a day. That means you could be called into work at any time, especially in case of an emergency.

COLLEGE AND TRAINING

Most park rangers get a four-year college degree. Coursework might include forestry, wildlife management, environmental science, tourism management, and law enforcement.

After college, new park rangers will receive on-the-job training. They need to know the history of the area, the stories behind monuments, how to protect the land, and how to keep visitors safe. Other subjects that new park rangers will receive training for include wildlife safety, search and rescue missions, fire prevention, and more. Most parks will expect park rangers to have a drivers license so they can drive park vehicles. Rangers may need to retrain every few years to stay up to date.

OUTDOOR ADVICE

Park rangers need to be trained in first aid, **CPR**, and survival skills. You can pursue this training now and be prepared when you're officially ready for a park ranger job.

NATIONAL PARK SERVICE EMBLEM

The National Park Service (NPS) offers a training program for many new rangers. It includes online courses and in-person training for two years.

PARK LAW ENFORCEMENT

Park law enforcement focuses on the protection of natural **resources**, monuments, wildlife, and park visitors. Just like towns and cities, national parks have laws the visitors must follow. Park rangers can give tickets to, and even arrest, people who break laws. Rangers also manage large crowds, **investigate** parks accidents and crimes, and do regular safety checks of park buildings.

Park rangers are also emergency responders. They react quickly when accidents and crimes occur. A ranger may need to give an injured hiker first aid. They join and even lead search and rescue missions. These mission can be very dangerous, but it's part of the job!

Park rangers rescue a trapped hiker.

TEACHING AND INTERPRETING

Many park rangers are teachers. They are often referred to as interpreters. A language interpreter hears one language and repeats the meaning of the words in another language. This is called translating. Park ranger interpreters are translators too. Instead of translating languages, they translate the meanings behind mountains, rivers, wildlife, history, and more.

Interpreters have a long list of duties. They lead guided tours. They show visitors where the bathrooms are. They explain park laws and safety rules. Some give educational **demonstrations** and answer questions. They lead **volunteers** in cleaning up the park or managing invasive species. Rangers are often tasked with training and overseeing new rangers.

Some rangers serve as actors! They help teach visitors about the history of the park though entertaining presentations.

CONSERVATION AND STEWARDSHIP

In addition to law enforcement and teaching, park rangers are also trained in conservation and stewardship. Conservation is the protection of natural resources, including water, air, soil, plants, and animals. This also includes the protection of cultural resources and historic landforms. Stewardship is the is careful management of something. This means that park rangers are managers, or caretakers, of our national parks' wildlife, resources, and monuments.

Rangers make sure visitors follow rules to keep parks safe. They sometimes work with scientists who are researching animal populations. Some may also write scientific reports based on their careful observations and conservation knowledge.

OUTDOOR ADVICE

Rangers are trained to fight wildfires. However, they're also trained to start controlled burns in areas that have become overgrown and where a wildfire could start.

Park rangers work with different kinds of scientists to help keep our parks pollution free.

MAINTENANCE AND UPKEEP

There is a less exciting side to a job as a park ranger. Rangers are often expected to help maintain the park grounds. This could mean something different every day. Rangers often clean up garbage left in campgrounds. They may need to clear fallen trees and branches from park trails. They plant flowers in spring, mow lawns in summer, rake leaves in fall, and shove sidewalks in winter.

OUTDOOR ADVICE

Did you know that the term "ranger" means to roam freely? Can you see why this type of worker was given that title?

Maintaining a clean, beautiful park is an ongoing effort. It sometimes leads to law enforcement duties. This is the case when visitors write **graffiti** on monuments or carve their initials into trees.

Rangers may also need to repair buildings on park land.

FAMOUS PARK RANGERS

Harry Yount was a soldier, hunter, trapper, miner, and mountain climber. Around 1872, he became a guide for scientists studying the Rocky Mountains. In 1880, Yount became the first gamekeeper for Yellowstone National Park. Yount told the U.S. government a ranger force was needed to help protect wildlife. Many consider him the first park ranger.

HARRY YOUNT

Johnson starred in a series of documentaries called *The National Parks: America's Best Idea*. In 2009, John viewed the documentary at the White House with President Barak Obama.

Shelton Johnson has been a Yosemite National Park ranger for over 30 years. He has spent much of his life teaching others about Yosemite and **advocating** for more diversity in U.S. national parks. Johnson is an expert on the Buffalo Soldiers, a Black **regiment** in the U.S. Army from 1866 to 1951.

THE RIGHT JOB FOR ME

Pay for park rangers often depends on the location and specific duties. Rangers with a college degree and other special training may receive more pay to start. Pay can be different from state to state. Federal jobs often pay better than state or local jobs, and offer more benefits. The average pay for park rangers in 2024 was about $70,000.

Park rangers are busy workers! However, it's a job that pays well, and it's a job that's perfect for people who love being outside. It's also a great job for people who like to learn new things. Does this sound like a good fit for you?

For some rangers, the best part of the job is teaching young people.

GLOSSARY

advocate: To argue for or support a cause or policy.

communication: The use of words, sounds, signs, or behaviors to convey ideas, thoughts, and feelings.

CPR: A procedure used to restore normal breathing when the heart stops beating. Short for cardiopulmonary resuscitation.

demonstration: An event in which people gather together to watch something being done.

environmentalist: A person concerned about environmental quality and especially with controlling pollution.

graffiti: Usually unlawful writing or drawing on a public surface.

invasive species: Plants or animals from one area that spread quickly in a new area and harm native plants and animals.

investigate: To study by close examination and systematic inquiry.

recreation: Having to do with activities enjoyed outside of work.

regiment: A military unit usually consisting of a number of smaller groups called battalions.

resources: Something that can be used.

volunteer: Someone who does something because they want to help and not for money.

FOR MORE INFORMATION

BOOKS

Allan, John. *The Outdoor Survival Guide for Kids*. London, UK: Beetle Books, 2025.

Walton, Kathryn. *Visit Yellowstone National Park!* Buffalo, NY: Enslow Publishing, 2024.

WEBSITES

Becoming a Junior Ranger
www.nps.gov/kids/become-a-junior-ranger.htm
The National Park Service website has a wealth of information for kids interested in becoming a park ranger.

Shelton Johnson: Park Ranger and Advocate for Diversity in the National Parks
www.metroparks.com/shelton-johnson-park-ranger-and-advocate-for-diversity-in-the-national-parks
Read a short biography of legendary park ranger Shelton Johnson and learn more about his efforts to make our national parks more diverse.

Publisher's note to educators and parents: Our editors have carefully reviewed these websites to ensure that they are suitable for students. Many websites change frequently, however, and we cannot guarantee that a site's future contents will continue to meet our high standards of quality and educational value. Be advised that students should be closely supervised whenever they access the internet.

INDEX